THE LAST PROPHET
by MAX

THE LAST PROPHET

Originally written in January of 2011. Copyright © 2013 Max Gold

All rights reserved. The onus is on the reader to attain a fully detailed and comprehensive cognizance of all the rights. No part of this book may be reprinted in any form, without the express consent of either the author or publisher. This book burns longest, for no discernable reason, if set ablaze on page thirty-three. For best results, douse the page in an accelerant such as a distilled spirit or gasoline. If neither is readily available, good old fashioned matches and a bit of patience should suffice.

This book has not been approved by The Council of Al-Azhar Al-Sharief Islamic Research Academy. This book has not been approved by The Supreme Islamic Shiite Council of Lebanon.

The publisher does not hold or express any of the ideas in this book. The publisher doesn't really give a shit about anything in this book. The publisher cannot be held responsible for any of the ideas in this book.

Max uses Microsoft Word, Quark XPress 8.0, Bic Reaction™ and Paper Mate® gel pens.

Front Cover Atwork: Angus Die by Francesco de Zurbaran
Back Cover Artwork: Wolf Eyes by fullmetal-fox.deviantart.com

LCCCN: 69-6969-69696

Printed in Saudi Arabia

ALSO by MAX GOLD:

"DON'T JUDGE a DICK by ITS FORESKIN"
God, Life & Revolution

dont-judge-a-dick-by-its-foreskin.com

&

"THE LAST TEMPTATION of MAX GOLD"

the-last-temptation-of-max-gold.com

REGRETS

To all of the believers of Muhammed's religion who are just trying to get by like most of us, I am sorry that this book will offend you; it's not intended for you. More importantly, to the believers of Muhammed's religion who have shown me kindness, good laughs and conversations... because of you I regret writing this book. I can only hope you understand who and what I'm going after with it. Please continue being kind and open; people need this.

DEDICATION

This book is dedicated to the promulgators and financiers of radical Islam, militant Islam, Islamic theocracy, Sharia law, terrorism and any ideology rooted in oppressing people to appease any kind of Santa Clause. While not one person, living or dead, is without fault, you are without question an embarrassment to humanity.

Bullets and bombs cannot counter your ideas and beliefs effectively; they only embolden them. Only an idea can battle another idea. You are empowered and can manipulate others with your ideas only through a fervent belief in a delusion; the hereafter.

While this book is a puerile, flippant assault to all (which in reality is very little) that you hold dear, it is not intended to oppress or destroy human life. While you are free to call for the downfall of Western civilization, the submission of others to your will, the destruction of a nation and liken those of other faiths to pigs and apes... you are not free to call my words hateful. My words are a story. My words are my free will. My free will is my purpose to life. You can't stop my purpose to life; my words. Now let's get on with it and word to your mother.

THANK YOU

Luke... Sammy, Jesse, Robert & Motorboak, Paul A. Motherfucking Fler
.

SCENES

ACT I...... pg. 10

ACT II...... pg. 12

ACT III...... pg. 16

ACT IV...... pg. 18

ACT V...... pg. 52

ACT X...... pg. 62

THE END...... pg. 63

MAX GOLD

ACT I

THE LAST PROPHET

MAX GOLD

ACT II

(Max Gold is in his kitchen preparing supper)

I'm making a Caesar salad on the Upper East on the eve of my thirtieth birthday. My favourite salad with a rack of lamb and creamed spinach; that's a meal. Everyone on death row opts for a steak for their last meal. Me, I'd go for lamb all the way.

While squeezing half a lemon, I weigh in on how I should be eating at a friend's place, but how that's just not quite appropriate now on account of her husband's recent death. He was a good man. Shit, he was one of the best men I've had the good fortune to come across. He succumbed to alcoholism and a nervous breakdown; before the Monday morning Wall St. heart attack.

Instead, I'm alone with a bag of groceries, two bottles of wine I wouldn't serve to company, and well on my way to becoming a puddle before midnight. Which, come to think of it, isn't such a bad thing.

Throwing a spoon of Dijon in a bowl, random thoughts on New York come to mind:

"Mr. Cab Driver, Lenny Kravitz was wrong. If I was you, driving down Lex and 165th, at any time of day, I don't care if it's Barack Obama trying to hail me down... peddle to the metal baby.

If you have to take a train, subway, or drive through a tunnel or span a bridge please... please don't say you live in New York. If you do live in New York City, start saying you live in Manhattan; this should end the confusion and stifle the posers.

Williamsburg isn't that fucking cool. It has a few awesome bars and restaurants. Apart from them, it's Detroit with a bunch of ambigay hipsters dwelling in converted lofts and apartments. And it's a $15-$20 cab ride one-way from the Lower East. Seeing as how the people who live in Williamsburg drive bicycles, cab fare doesn't concern them.

You know someone works on Wall St. when they don't even hesitate to crack open an 8 year-old bottle of Nappa Cab... to cook with.

(*MAX GOLD adds a tablespoon of Worchester, salt and pepper*)...

Today, a homeless man singing "What a wonderful world" paused the great cover version he had going to ask me for $5. The Federal Reserve has got to get inflation under control. I gave him the address to a homeless shelter/soup kitchen instead. He looked at me in return as if I were on what he was trying buy.

Women here are, by and large, a better breed; they're just better looking and better turned out. The clothing they're wearing now will be what women in most other cities will be wearing in six months to two years from now.

This is probably the only city where you don't notice gay people; they seem to just blend in. Anywhere else... when you see a guy wearing a black and white striped shirt with a scarf wrapped around his neck you reflexively think "cocksucker"; not in a malicious way, just as a kind of note to self.

(*MAX GOLD adds an egg yolk*)...

Being able to hit a up a deli at 4AM rules.

(*MAX GOLD adds another egg yolk*)...

Being able to buy a pack of Guinness at that deli at 4:03AM can help you get laid.

(*MAX GOLD adds just a dab of hot sauce*)...

Sure this shithole is expensive. So is a hot rental.

(*MAX GOLD stirs in the garlic*)...

Previously leading up to 9/11 you had to live here for 10 years before you could constitute yourself a New Yorker; that, or you had to have had a couch found on garbage day as a piece of your living room furniture. Now, according to me, unless you lived here on September 10th or before then, you're not a New Yorker. The New Yorker club is over... closed... no longer taking membership. This city, while still awesome, is different. Like an athlete coming off a second concussion, it's back in the game... but all too aware of the past hits and the potential one it might not see coming. At any time.

(*MAX GOLD throws in the anchovies*)...

All the things that make me hate this town... the horns, construction, the lonely and stressed out people murmuring to themselves, the people struggling and hustling to get by, the pricey tabs, the cell phones, the obnoxious accents, the self-obsession and the rude ingrates... are all of the things that make this joint so endearing."

(*MAX GOLD mixes in the romaine with the parmigiano reggiano and croutons*)...

And then I start ruminating about the kids. Not because they're particularly dear to me, though deep down somewhere they are, but because I can't stop noticing the drawing of a smiling sun one of them drew for me. I hope they don't share my experiences.

They should hope so too.

ACT III

(MAX GOLD is sitting on the living room couch watching a television documentary)

Four or five glasses in, the wine I'm drinking suddenly becomes fantastic. And the flick on Lombardi is getting the job done as well. I'd play hard every play for a man who cared that much. It's easy to follow when you think you know where you're being led.

(MAX GOLD gets off the couch and heads for the washroom)

MAX GOLD

ACT IV

(MAX GOLD *is standing in the washroom*)

Washing my hands afterwards I felt the ground begin to shake. The lights over the mirror began to blinker. The walls shook violently. The soap fell off of the sink, the hoop on the bowl flapped up and down and all of the shit on the shower ledges fell in the tub. Finally the entire apartment went black and stillness replaced the chaos. As the lights came back on, sitting on the ledge of the bath was a man. I screamed. He just sat there unfazed, with a wry, shit eating grin on his face. He was dressed in a black suit, black tie, white shirt, ferragamo shoes and a black cashmere coat. He had an olive complexion to him and was no less than six foot five.

MG: What the...

UNKNOWN STRANGER IN THE WASHROOM: I've been waiting a long time to get here. Ride in was turbulent, but I've done worse.

MG: (*mortified*) Lucifer!

LUCIFER: (*in an offended tone*) What? Are you reading my fucking birth certificate? Call me Satan.

MG: This is impossible, I'm hallucinating. I have food poisoning. A man just appeared out of thin air right next to me!

SATAN: Man? Kid, if an archangel were to manifest in his true form, he wouldn't fit in this room. I'm in human form to make it easier for you to handle the whole situation.

MG: (*feeling faint and gasping for air*) I'm, I'm...

SATAN: You're flummoxed. It's ok kid. Take a slow and deliberate breath. You're just in revelation shock. Normally this sort of thing is done in the wilderness; the open skies and fresh air have a sedating affect.

MG: So why... (*gaining composure*)... why are you doing this in a washroom?

SATAN: This is going to sound really cliché, but we don't have a lot of time.

MG: For what?

SATAN: To save humanity and evolution.

MG: (*in a hurried cadence*) How can you save humanity? How can you save evolution... that's an oxymoron; evolution preserves itself through natural selection and, in turn, natural selection saves humanity. Saving humanity or evolution would be defeating the purpose of both; they need to save themselves as a function of their very being.

SATAN: Kid, I love what I'm hearing. god would be very displeased with you.

MG: Then why did god give me a brain with which to displease him? Why give the choice? Isn't choice a temptation? Aren't you, after all, meant to be the great tempter?

SATAN: Now that's the spirit!

MG: Good for me. Either tell me why you're here or I'm checking myself into the nearest psyche ward.

SATAN: Tuck in your tampon. I just traversed thousands of universes and over six-hundred dimensions to get here and this is the welcome I'm met with? You have a valid question and I'll answer it. All I ask for is a little civility in return.

MG: Sure, fine. Whatever.

SATAN: Humanity, like all things in existence, is in a constant state of evolution. Every facet of humanity is in a constant state of progress; be it physical, social, technological, ideological, or what have you. The problem is the little blip of humanity going on here is doomed. Was from the get go.

MG: How so?

SATAN: Just because something is inherently designed to evolve does not necessarily mean it will meet out its evolutionary potential. Sometimes help is required, but... (*MAX GOLD interjects*)

MG: The road to hell is paved with good intentions?

SATAN: (*laughing*) And the road to heaven isn't?

MG: What kind of help are you referring to?

SATAN: About ten thousand years ago we saw humanity stumbling around with itself. Generally fucking things up and making no progress. Some of us wanted to keep our hands out of it, some of us wanted to improve things for amusement, and some of us actually wanted to help you out. So we... (*cut off again*)

MG: Who is we?

SATAN: The most powerful conscious entities in the closest proximity to your universe.

MG: What are you?

SATAN: Hard to say. We consider ourselves a species. Humans have always called us angels, spirits, gods, extra terrestrials, ghosts, or any title you'd ascribe to anything mysterious or unseen. We're what humanity might be in a few million years; though infinitely more experienced.

MG: Did you or god create this universe?

SATAN: Hell no. god is no more powerful than the rest of us, but I'll explain that later. We do create universes from time to time, but solely for the purpose of resource discovery and experiment, and we always ensure they can't foster sentient life. When we found this place it was like if you were to take a walk in a forest and find a tree with a nest in it. But I'm getting sidetracked here kid.

MG: Sorry, you were saying...

SATAN: Yeah, so humanity was at a standstill. We knew that we could help you out. I wanted a more indirect, hands off approach to things; a hand-up, not a hand-out. god wanted a more forward intervention.

MG: So what did you guys do?

SATAN: We went with my approach. Slowly, we guided a few people, subliminally leading them to discover and build on things like arithmetic, tool making, irrigation and concepts like the rule of law and government.

MG: How did you "guide" these people?

SATAN: Basically implanting the ideas and concepts in certain people's minds through their dreams. The ideas would then present themselves in waking life when the person we implanted was faced with a problem that the implanted idea could resolve.

MG: That sounds heavy.

SATAN: It was. But it was an unfettered success. Some religion materialized as a side effect, but only as means to cope with and understand the advanced new ideas – not as a directive to instruct these new ideas. god saw the formation and concept of religion as an extremely powerful tool. I saw it as a dangerous one.

MG: What is religion? I mean why have we always had them?

SATAN: Religion is an organized way of reconciling the individual with the outer world. It's a way for primitive consciousness to make sense of the harshness of reality and to try to come into harmony with it.

MG: But if religion is a way to make sense of reality, why all of the gods, angels, spirits and other sensational characters?

SATAN: The gods and the like are used to symbolize human traits and ideals; they give the mind something it can identify with. They also give the mind things to strive for and be like, or avoid acting like.

MG: So you're saying all of the duality in most religions is really just a mirror held up revealing our duality; our good and bad.

SATAN: Precisely. The Hindu's understood this with their concept of the trinity; the creator, preserver and destroyer. The Chinese got it as well with the concept of the yin and yang; two opposites simply comprising two pieces of the same thing. Thousands of years ago language was primitive. Symbols were the only way for people to really understand things. Symbols attributed to humans or human-like entities were understood best. In your present time, this is no longer the case; words are what people get, not the symbols. This is why so many people take religions "word for word" when they were intended to be interpreted through the symbols.

MG: But what about the Aztecs who sacrificed people to rain gods? Where's the symbolism there?

SATAN: Well, first off the Aztecs lived in harsh conditions and their primitive minds needed some rationale for the lack of rain, so they make up a rain god to, as I mentioned to you before, make sense of it all. That rain god would be the Aztecs ideal of a spoiled and sadistic tyrant; the ideal of an uncompassionate human. Of course there is no rain god who needs humans murdered before he'll provide life giving water in exchange. That's retarded. But so is primitive consciousness. Where the idea to kill or sacrifice people comes from... that's equal parts sadism and psychosis.

MG: Satan, do you mind, I need to go to the washroom here for a sec?

SATAN: Oh no, go for it. (*Satan heads out to the kitchen and pours himself a glass of wine. When he returns he knocks on the door*)

MG: Hold on. (*Satan takes a sip of wine*)

SATAN: This wine tastes like... I'd rather drink white. god it's fucking awful.

MG: I got it on sale. $5.95 a bottle. It works.

SATAN: Point taken.

MG: (*in a tone suggestive of an impending question*) Satan...

SATAN: Shoot.

MG: Why are humans so much further evolved than other species on this planet? I mean, crocodiles have been around since the dinosaurs, but they haven't developed speech, irrigation, math and, well, you know what I mean.

SATAN: (*smiling*)... Great observation. It's the same observation that's given humanity a divine complex; one that posits it as being above everything else on this planet.

MG: Yeah, like we're the only thing here that counts.

SATAN: Well you are the only thing that counts. Everything else, while sentient and deserving of fair treatment and compassion, is too stupid to tell the difference. Animals can be domesticated and can give and receive love, but that's only because they've been taken out of their natural environment where they'd be receiving and giving love to their own kind.

MG: While I'd like to, I really can't argue with that. Fair statement.

SATAN: Of course it is. But to answer your question, humanity's advanced evolution is just an anomaly. A gift really. It's extremely hard to think that somewhere down the line a few monkey's mutated and branched off to form a new species, but that's how it works.

(Satan begrudgingly takes a sip of wine)

A Great White Shark is a fish right?

MG: Essentially, yes.

SATAN: Does it look like a fucking Gold Fish?

MG: No.

SATAN: That's physiological evolution. The difference between the Great White and humans is that the Great White had its last major evolutionary change way the fuck before you guys came on the scene.

(Satan sips again from his glass and swallows in further disgust)

I can't drink this shit.

MG: It's not that bad. It's French.

SATAN: It tastes like its French Canadian.

MG: You just teleported here, don't you have the power to make your own?

SATAN: I made a bet with Jesus once, much to that end.

MG: Really?

SATAN: He was partying and tripping balls out in the dessert with some friends and going off on how almighty he was. I bet him he couldn't turn a rock into a loaf of bread. And of course the guy backpedaled and stalled with some bullshit about how it wouldn't be right because I was asking for a display of his powers, instead of being shown them.

MG: But could he have?

SATAN: He thought he possessed the power to fly. Come to think of it, it's too bad they weren't on a balcony that night.

MG: If he wasn't powerful or special, why did you even bother challenging him?

SATAN: Those were interesting times. We knew it was a shift in humanity's progress. A message of passivity had to get out eventually; people were really making life difficult for one another. More organization was needed in the future. There were a lot of guys running around and upsetting the establishment. A buddy of mine told me about this one guy pissing a lot of people off in a cult with a saviour based prophecy, who had a good chance of leaving behind a strong cult of his own when he was executed; a cult that would then vie with other cults to form the first dominant religion. I visited him just to see what he was all about. It may have been wholly to do with the drugs he was on, but man was he ever delusional. When you think you're a child sent to a planet by the one being who created that planet... for the sake of saving everyone on that planet... by dying... even though everyone else will die eventually... that's good shit you're taking.

(Satan finishes his glass)

I felt compelled to call him out on his bullshit. He was just a human among humans.

(Satan produces a bottle of Caymus Special Selection out of thin air, followed by two glasses. He fills the glasses and hands one to MAX GOLD)

MG: *(in disbelief)* Holy shit that was awesome! Can you do that with pussy?

(Satan smiles)

MG: *(Taking a sip)* So why did Jesus stick? Why was he a part of the angels plan to advance the human species?

SATAN: Ask god. I was against it. I wanted to wait for the Roman Empire to fall before we tried to establish more advanced ideas. Most of the others were indifferent and stayed neutral. god won out by the majority of the willing minority.

MG: So you're saying that god isn't above any of the other angels?

SATAN: Kid, god is an angel. His name is god. In our language god means "a cool summer night's breeze." He's a real insecure and egotistical prick. Reckless, unscrupulous and selfish, we can't see eye to eye on anything and can't even bear the sight of one another.

MG: So why is god always the creator and benevolent figure in most, if not every, religion? And why are you the protagonist?

SATAN: Because I always show up too late. god is younger, stronger and faster than me. If it takes me two weeks to get somewhere, god can make in a few days. Whenever we've decided to intercede in your progress, god and his crew arrive first. And to the victors, go the spoils. god doesn't really give a shit about your well-being. He just wants to raise you up high enough so that your collective fall will be all the more harder. Humanity is just a board game to him. He really gets off on humanity thinking that he's the one it owes everything to. It's actually quite comical.

MG: So why don't the rest of you stand up to him?

SATAN: Most of the other angels have their own shit to deal with. My allies are outnumbered by god's. It's a struggle and in the grand scheme of things, you really don't mean that much.

MG: And why are you always the bad guy?

SATAN: Because god takes out his grudge on me by making the prophets he brainwashes afraid of me. He knows I'll eventually catch up and meet with them, but he makes sure they'll never hear a word of what I have to say; despite my being his senior. He tells them that I rejected his love and authority and that I hate humanity and blah, blah, blah. What's worse is that he teaches them that if they ever question him, they are siding with me; a stance that, by default, will land any sceptic in a furnace forever. I don't mind though. It's worth it for the symbolism and does make for a better read.

MG: What caused all of the animosity between the two of you?

SATAN: I'm smarter than him and he knows it.

MG: C'mon, that's it?

SATAN: And he walked in on his wife giving me a blowjob at a Christmas party one year. He retaliated by destroying my record collection and there's been enmity between us ever since.

MG: Hey, a blowjob is a pretty low...

SATAN: Gimme a break. She was giving blowjobs to everyone there like they were hors d'oeuvres. One blowjob is worth ruining four-million millennium's worth of music? No fucking way man!

MG: I suppose.

SATAN: Why do you think god is so against promiscuity! Everyone gets laid where we come from except for him and a few of his friends; one of whom I really need to talk to you about.

MG: Before that, how did Judaism come about?

SATAN: It was an elementary project. The Jews were a tribe, but unlike most tribes they just wouldn't settle down; they liked exploring and taking risks. Naturally, we were impressed. Then came Abraham. He was steeped in Egyptian culture and saw the power of structured religion. He thought it could do some good for his tribe.

MG: You mean to tell me that Abraham got ideas for Judaism from Egyptians?

SATAN: Not just ideas, whole concepts. Many of the myths and motifs in the old testament and other Jewish books share uncanny parallels with Sumerian mythology; which of course laid the foundation for most Egyptian beliefs. Noah's Arch is simply the Epic of Gilgamesh with heavy line edits. That's just scratching the surface.

MG: Holy shit.

(Satan pours two more glasses)

SATAN: I mean, Abraham's claim to fame was monotheism; that there is only one god. The Sumerians and Egyptians believed that as well. But they believed that the one god had many helpers; a support staff if you will. All Abraham did was label the support staff angels.

MG: But why the need to kick off Judaism?

SATAN: Because we liked their tenacity and work ethic. We thought it could set an example and possibly spread for humanity's benefit. It spread alright, but just not at all as intended. The whole thing was fucked from the beginning because of three fatal flaws.

MG: Which were?

SATAN: god fed all of that fear thy god crap to Abraham – it was intended as a joke, but it didn't play out as such. Then Abraham made up the saviour/messiah bullshit; which we tried to patch up with Jesus, only to result in disaster. I mean, how can one person make things right for everyone? It's impossible. You make peace in the world idea by idea, person by person; people need to save themselves. Lastly, Abraham labelled his tribe the chosen people. And then he promoted same-religion marriage, frowned upon conversion and made the whole fucking thing a little clique; something god admired. It was a display of arrogance that I would have only reserved for god, but nevertheless, there you have it. That proclamation of the chosen people alone likely fucked the Jews because they never had the desire or muscle to enforce such a statement.

MG: Forgive me if I construe all of that as laced with anti-Semitism.

SATAN: It's just fact. god's divisiveness shines through in all of our evolutionary experiments... you know, religions. Most religions do view people who don't adhere to their religion as inferior in some way. It's not exclusive to Judaism by any means. But I understand your concern.

(Satan exhales slowly and takes a deep breath, as if disappointed)

Listen kid, I can't explain everything for you. I can't edify you with the whole thing. We just don't have the time.

MG: Then why are you here?

SATAN: I'm here to inform you that human evolution, and humanity itself, is in serious trouble.

MG: Lemme guess... an asteroid is coming our way?

SATAN: Yes, but not for a long while yet.

MG: Oh sure, no big deal there. We're all going to be gasoline in the future because of a massive dust bowl, but, hey, not important right now. Are you crazy?

SATAN: Humanity's troubles are more imminent, relatively speaking.

MG: How so?

SATAN: Our last real intervention with humanity was the most botched undertaking we've ever undertaken. Christianity was spreading like herpes on a college campus in the middle of frosh. It was alarming, to say the least. We felt we needed to give you guys more options; too much of one idea always stifles out other possibilities and the chance for better ideas to come along and further progress.

MG: When was this?

SATAN: About 1400 years ago.

MG: I think I know where you're headed.

SATAN: Of course. I was adamant on disseminating some philosophical concepts to the people surrounding the core region of Christianity. They were good people, open minded people. They lived in peace, but were particularly vulnerable to the coming wave of Christianity. god wanted to give these people another religion. Me, I thought they were ready for logic and reason.

MG: So what happened?

SATAN: I got a head start on god and left our realm twenty-one days ahead of him. It would have allowed me to get here early enough to convince whomever he spoke with that he was full of shit.

MG: Did you get a flat tire or something? Why the hell was there a problem?

SATAN: A great tragedy occurred. A tragedy worse then had god shown up himself.

MG: Yeah...

SATAN: Fucking Gabriel. When god realized that I left for Earth he was too hung-over and coked-out to move. He didn't really care about what humanity was to be imparted with, just so long as what it got could be, by any and all means, turned into a religion. So he sent the fastest one of us all to beat me and get the job done for him. And he also sent out the worst of us.

MG: How so?

SATAN: Gabriel has to be the most repressed, self loathing, and sadistic angel I've ever come across.

MG: So why did god trust to send him?

SATAN: Because being a weak minded being, he'll do anything to curry favor with god; to gain acceptance with the in group.

MG: What did he do when he got here?

SATAN: He found the most dimly minded person that he could, to start a religion. I say start a religion because advancing humanity was the furthest thing from his mind; usually with god's formula the religion followed the core ideas. The tragedy of it all was hilarious at first.

MG: How so?

SATAN: Oh Kid... this is good. The human mind has a limited auditory memory. As such, we usually get the people we choose to contact directly, to write shit down. Well, Gabriel, being Gabriel, selected a guy who couldn't even write!

MG: You're speaking of Muhammad, right?

SATAN: Mark Twain, actually. Of course I'm speaking of Muhammad! Who else has ever claimed to be the scribe of a book written by the creator of the universe, as dictated by an angel? But it gets better.

MG: How?

SATAN: When Gabriel realized Muhammad was of no use, he panicked and made Muhammad promise that he would recite the revelation to someone who could write it down. It was such a hack job that Gabriel had to keep checking up on Muhammad for over twenty of your years just to make sure his friends were able to record most of the revelation. A simple revelation took over twenty years to complete!

MG: Did you try and help him? Surely you had enough time to intervene?

SATAN: When I confronted Muhammad in his cave, I missed Gabriel by about six days. He'd probably been smoking hash for three or four days; strong shit too, probably Lebanese. I tried to convince him of the danger of what he'd been told to write... and what his friends were adding themselves, but he would have none of it.

"No, they will believe, they must believe. I will make them submit" is all he kept mumbling.

MG: Wow.

SATAN: Oh yeah, he was out of his fucking gourd. When Gabriel told god what he'd done, expecting to hear praise, god lost it on him. Even he knew he screwed up by not overseeing the revelation himself.

MG: Did god try and intervene? He'd have to be able to step in, wouldn't he?

SATAN: After he reprimanded Gabriel, god only spoke two more words of humanity. And they were his last.

MG: Which were?

SATAN: "They're fucked."

MG: That's it? He just gave up?

SATAN: I'm afraid so. He tended to his own affairs from then on in; as he should have from the start, given his lack of sincere interest in you guys. Like I mentioned to you, Gabriel tried to make things right by constantly visiting Muhammad to ensure the revelation was documented properly... but to no avail.

MG: So why do you care?

SATAN: Kid, you've forgotten already. It's not about me, or us. It's about you. We care because we see humanity's potential. We even see a bit of you in ourselves. We're not going to be around forever. Something has to fill our shoes. Hopefully you can fill that void. If not, there are others. Humanity is not bound to this planet. There's more out there – a lot more. And to get there you can't be content with scraping by here; toiling in stunted, stifled states of mediocrity.

(Satan pours more wine for the two of them. Satan inadvertently spills a drop on MAX GOLD'S pants)

Oh, I apologize for that.

MG: There's a Tide-Stick around here somewhere.

(MAX GOLD rummages through the washroom medicine cabinet and finds it. A minute later the stain is removed)

SATAN: I've gotta get one of those.

MG: Yeah, they're great aren't they? You can get them in any grocery store check-out line or pinned up in most laundry detergent aisles. So why didn't you guys try and counter Muhammad's religion with something else?

SATAN: By then it was too late. If you meddle with something like the collective thought processes of a species too much and for too long, it just results in chaos. So I stuck with the initial method.

MG: Implanting ideas.

SATAN: Exactly. Just on a smaller scale than we did before.

MG: What was your approach?

SATAN: The Renaissance. I helped start it.

MG: What do you mean by helped?

SATAN: With the aid of some old friends I helped bring about a creative rebellion, or counter culture, to the Dark Age. When the Renaissance started winding down, I conducted my masterpiece.

MG: The enlightenment. That's why your name means the bringer of light; as in to enlighten. (*MAX GOLD takes a large sip of wine*) Shit, I've gotta ask you while you're here, did Adam and Eve really exist?

SATAN: Nope.

MG: (*slamming his glass on the sink and breaking it*) Whoops... sorry about that... Motherfucker, I knew it!

SATAN: You got your start as a bunch of monkeys, kid. If it's of any consolation to you... so did my species. Adam and Eve came about in religions for the simple reason that every story needs a beginning. They're played out in a whole smorgasbord of myths. In some cultures Eve is seen as wise for listening to the tempter because she finally does realize that she is indeed naked, or she discovers or dares to uncover an unknown truth of some kind; her brain starts working, if you will, and she leads the man forward with her.

MG: How did you start the enlightenment?

SATAN: Kid, with what I love the most... Music. First I helped inspire the greatest musicians in your history. Today there are three to six people playing instruments in your orchestras. At the start I had guys writing parts for a hundred, sometimes close to two hundred instruments; and the songs could be two or three hours long. But anyway, I then helped take philosophy, medicine and the social sciences to another level. I didn't write the notes, invent the concepts, conduct the procedures, or dictate the words. I just helped inspire people to find it within themselves to follow their love, their passion.

MG: It worked. Humanity took some extreme leaps in thought during the Enlightenment. The standard of living in the West really took off.

SATAN: It did. And it's starting to spread everywhere now, albeit slowly. Things are looking up. There will be struggles, but there is enough intelligence here now to persevere through anything.

MG: Wanna do a shot?

SATAN: I thought you'd never ask. Jager?

MG: Lemme see what they've got here.

(Max Gold heads to the liquor cabinet and returns with a bottle)

It's only room temp.

SATAN: It's not a bowl of soup, who gives a shit?

(Max Gold pours two shots and they drink)

MG & SATAN: Cheers.

MG: So you and your buddies pulled off the Renaissance and you managed to orchestrate the Enlightenment. Good going. What's the problem?

SATAN: Muhammad's religion.

MG: It will adapt to modernity one day.

SATAN: Oh, many adherents to it do understand the Enlightenment, or at least the core ideals born from it. The problem lies with those who don't, never will, and refuse to.

MG: Sure, but the hard-line followers will always be there, and they're stuck in lands with little influence. Who gives a shit about a bunch of nutjobs who enforce the length of a beard or fashion styles in parts of the world that don't affect me?

SATAN: Well, that's where you're wrong. Wrong and naive. These nutjobs you're referring to, let's call them radicals...

MG: Sure.

SATAN: You know what, let's do a Jagerbomb.

MG: (*belching*) Fuck, why not?

(*Satan pours and lines up two Jagerbombs*)

MG & SATAN: Cheers.

SATAN: Where were we?

MG: Nutjobs.

SATAN: Yes, the radicals. Well, they take Muhammad's words as a law much the same as an engineer takes gravity as a law when planning to build something. The problem is, you can't build much with Muhammad's laws. Understanding them and observing them, with any real commitment, leaves no room for progress and opportunities.

MG: Come on. Muhammad's religion is peaceful.

SATAN: Actually, it isn't. But we'll get to a minor exegesis of his book in a bit.

MG: What's an EXuHguHSis?

SATAN: It's a critical explanation and interpretation of a religious text; something which Muhammad's religion forbids in the case of its own book.

MG: You can't even analyze Muhammad's book?

SATAN: Nope. It's perfect, remember. Never been altered, totally original. 100% god's word, as dictated by Gabriel for over twenty years to an illiterate who had his friends write down what he remembered. Questioning it, or trying to change it, is punishable by death. And whoever kills the questioning blasphemer gets a first class ticket to paradise and an orgy with sexually inexperienced women.

MG: That sounds harsh.

SATAN: I know. To think you should die for asking a question or being slightly critical of a... (*MAX GOLD interrupts*)

MG: No, I mean the virgin part. There's nothing worse than a lame lay. It's a good thing this paradise or heaven doesn't exist. I mean, it doesn't exist, does it?

SATAN: Kid, once you're planted, and the last scoop of dirt is thrown on your plot, and the spade is tapped down a couple of times... it's over... you're not going any further or farther. If humanity understood this, you'd all be a lot more productive and kinder. But we're wasting time here kid.

MG: Yeah, back to the radicals.

SATAN: They want one of two things. Either the submission of the Western world to Muhammad's religion, or, if that's not attainable, the destruction of it would be just fine.

MG: How do they go about doing this?

SATAN: There are two schools of thought. They can try and populate the

Western hemisphere and gradually convert people, or gain enough influence in Western lands to then impose Muhammad's religion on the population. The really crazy radicals just want to bring the whole thing down.

MG: But why? Why don't they just worry about themselves? I mean, get a fucking life, or a library card, and find something else to do with their lives.

SATAN: Ugh, because.

MG: Because why?

SATAN: Because Muhammad's book says so. Kid, when you look closely at Muhammad's religion and the book he kind of wrote... you'll see how any association with either is fundamentally incompatible with humanity and its respective evolution.

MG: Another shot?

SATAN: Absolutely. Let's do a grenade.

MG: What's a grenade?

SATAN: You pour two shots of Jager and place them in the rock glass. When you're ready to drink, you pull one shot out of the glass and the other shot drops in the Red Bull. You drink the shot and then do the Jagerbomb. Pulling the first shot out of the glass is like pulling the pin on a grenade... get it?

MG: That's fucking genius!

(*Satan pours the drinks this time*)

MG & SATAN: Cheers.

(*They drink the first shot*)

MG: (*wincing a bit*) Fuck me.

Satan: I know, eh?

(*They both reach for their Jagerbombs*)

MG & SATAN: Cheers.

(*They drink their Jagerbombs*)

MG: So how did so many people convert to Muhammad's religion?

SATAN: Like any of the major religions with an agenda, through fear and violence, that's how. It took Christianity a good three-hundred years or so to use mass-violence to convert non-believers. Muhammad got down to it right off the bat. You were either with him or against him. And if you were against him, your tribe was pillaged. If you didn't convert, you died. Christianity played this game as well, but, again, it wasn't established with violence; violence took it to another level well after it had a voluntary following.

MG: Sounds like you're almost defending Christianity?

SATAN: Oh god no. Christianity has its terrible legacy. I mean to say that the central figure of the Christian faith was not a violent man and never forcibly converted a person who didn't believe what he did. Christians can't openly wage a war in the name of Jesus; he never set that kind of example. But people can openly wage a war in the name of Muhammad; he set that example with his actions and even made it an obligation, under certain circumstances, in his book; well, the book Gabriel told him to write. And the same book he had his friends write for him.

MG: When you look at it like that, it's all pretty fucked up. Why is Muhammad's religion so popular then?

SATAN: Ah, for a few reasons. It states that Judgement Day, a Christian concept derived from faiths before it, can't take place until the end of the world or until everyone here believes in Muhammad's religion; this motivates believers to convert people. Many people are converted out of fear and threats of violence to this very day. Muhammad's religion also promotes polygamy, many of his believers live in poor socioeconomic regions, and, naturally, birth control is a no-no.

MG: Hey, but Christianity got its numbers more or less the same way. It makes sense.

SATAN: True, but you're forgetting a few important points, kid. Christianity had a four-hundred lead on Muhammad's religion – and both have the same numbers today. That, and Christianity is slowly dying off; educated people are having less children, because educated people realize you have a better life when you're feeding your lifestyle and savings and not several mouths. Muhammed's religion has only just begun to grow.

MG: Well now you just sound like a...

SATAN: A Muhammad's religionphobe?

MG: Yes.

SATAN: I'm not afraid of Muhammad's religion for an irrational reason, kid. I'm afraid of all religion, but with Muhammad's there are timely and relevant reasons for concern; as I will reveal to you in a moment. But first, let's have another drink.

(Satan brandishes another bottle of Caymus Special Selection from thin air)

You look too drunk though.

(Satan closes his eyes for a moment)

MG: Holy shit! I'm sober! How'd you do that!

SATAN: I filtered your blood. Don't ask me how. I don't have the time to explain, and since you lack a degree in medicine and quantum physics you'd never get it anyways. Just be glad that you can comprehend me coherently and actually taste this next glass.

MG: Thanks.

SATAN: Any time.

MG: So where were we?

SATAN: Well, we've talked about how a great way to ensure your religion grows is by killing people won't join it. But guess what Muhammad's book says to do to the people who stop believing in his religion? I'll spare you the suspense; it says to kill them as well.

(Satan holds his glass on an angle and fixates on it)

Remember how you thought I was a Muhammad's religionphobe?

MG: Yeah.

SATAN: Well in lands that are Muhammad's land, if you don't believe in his religion, you can't partake in any civil or federal policymaking. You're given second class status. You may or may not be able to travel in certain areas and you may or may not be able to enter certain places. You are not equal, by law and custom. You are not equal unless you believe in Muhammad's religion.

MG: That's the complete opposite of liberty.

SATAN: Exactly. Am I a bad angel for being concerned about a religion that creates such a condition, and which compels its believers to accept such conditions?

MG: Debateable.

SATAN: How can you preach harmony when you don't cooperate with others who have differences? You can't.

MG: Why do believers in Muhammad have so many different, often repressive, standards when it comes to women?

SATAN: That's all Gabriel. He puts the "gay" in Gabriel.

MG: Gabriel is gay?

SATAN: Gabriel couldn't get hard for a woman if god asked him to. He's a misogynist of the worst stripe.

MG: No shit?

SATAN: Being gay is natural if you're... gay, and that's totally fine, but it takes a repressed homosexual to hate a women so much as to force her to wear a bag over her head just to avoid having to even think about sexual relations with her. And it takes a special kind of idiot to take pride in wearing a bag or curtain over her head to demonstrate her pride and choice in supporting a custom of sexual repression.

MG: You've got a point there.

SATAN: Kid, according to Muhammad's religion, a woman's testimony in court is worth two thirds less than a man.

MG: While that's horrible, it's not like women over hear have been equal for that long. Women have always been controlled, submitted and treated as lesser beings.

SATAN: You're right. No, you're absolutely right. But the concept of women eventually getting rights was built into Western thought. The seeds for this concept just aren't there in Muhammad's book. We'll talk more about how his book is immutable in just a second. But, look, when you have a book that states you need to be kind to animals, but that it is permissible to beat a woman if she steps out of line with a man's wishes... it's a problem. The lack of education provided to many young people being groomed to be believers of Muhammad's religion is a problem. And it's a tragedy. If your whole scope and outlook in life revolves around a 1,400 year-old book that proclaims it is the only truth... what hope do you have for the future? How can you adapt to the world as it changes every day?

MG: Can't multiculturalism weed out all of the negative aspects of Muhammad's religion? Can't multiculturalism help people from other cultures adapt to Western culture?

SATAN: Nope. Western culture, which is multicultural, let's people be free to do what they wish within the law. It does not give people the right to dictate what Western culture must be, or the right to pick, choose, or have their own laws that they are comfortable with. The goal is to adapt to and embrace a way of Western life, without holding on to the life you came from.

(*Satan lights a cigar*)

If I wanted to move to move my family to China...

MG: You have a family?

SATAN: Hell no. Hypothetically, if I had one though, and wanted to move it to, say, China, ok, great. I'd move there in search of work and a more stable life. But how successful would my transition be if I expected China to lose its identity to accommodate me? How successful would China be? How successful would I be if I refused to embrace the Chinese?

MG: What do you mean?

SATAN: If I go there, I better know what I'm getting into. I'd be nuts to think businesses are going to have literature and signs translated for me; but if they were smart, they'd take the initiative to gain my business. I'd be nuts to expect welfare payments and free translators and ethnic transition councillors. I'd be nuts to take offense to any Chinese customs or traditions. I'd be nuts to expect the Chinese people and Government to care about the issues of the country I immigrated from. I better know that I'm going there to become Chinese.

MG: What do you mean?

SATAN: If I go there, I better know what I'm getting into. I'd be nuts to think businesses are going to have literature and signs translated for me; but if they were smart, they'd take the initiative to gain my business. I'd be nuts to expect welfare payments and free translators and ethnic transition councillors. I'd be nuts to take offense to any Chinese customs or traditions. I'd be nuts to expect the Chinese people and Government to care about the issues of the country I immigrated from. I better know that I'm going there to become Chinese.

(Taking a puff of his cigar then staring at it)

Multiculturalism is a failed idea. It's the notion of good people with good intentions, but it's a disaster.

MG: Why?

SATAN: Because it fosters the notion that it is fine, and even healthy, to hold on to your old life. And, as such, you get newcomers who have a hard time adapting to their new culture. They don't evolve because they hold on to their old ways. See kid, not that you'd expect many traffic signs in English or Government assistance in a place like China, but their way is pretty simple: come if you'd like and if you have something to offer. But if you can't make it on your own and do it our way, try somewhere else – no one forced you here. Ever heard the saying "When in Rome do as Muhammad does"?

MG: But what about the line in that poem... "give us your poor, give us your sick..."?

SATAN: Oh sure, we'll take them. But they better be willing to work or they better have a family member to pay the hydro.

MG: Surely time will help smooth things over and future generations will care less about Muhammad's religion, even if they were born into having to believe it?

SATAN: Let's hope so. But a lot of damage can be done before that. As long as terrorists and ideologues who plot to harm Western civilization keep gaining momentum, this vague period of time to smooth things over is a long way off.

MG: Gaining momentum for what exactly?

SATAN: Haven't you been paying attention? Muhammad's religion must, according to its more adamant adherents, span the entire globe, or the end of the world must take place before you can all be judged and get out of here. To get to that point Muhammad's religion needs to spread its influence or start a biblical war.

MG: But don't fanatical Muhammad's religion believers see us as waging the biblical wars?

SATAN: No. They see Western countries as falling for their trap. It is they who have provoked an attempt to start a biblical, or holy, war.

MG: How so?

SATAN: 9/11.

MG: Are you kidding me?

SATAN: No. They were ingenious with that one. For the cost of a small number of airline tickets, they managed to declare war on Western civilization. And they managed to hurt it a great deal. That was the day Muhammad's religion declared war on the West.

MG: You're saying the people behind that really saw themselves as playing a part in a grand scheme?

SATAN: That's exactly what I'm saying. They believe they are doing god's work.

MG: C'mon, those attacks were over foreign policy.

SATAN: Do you think any of those guys gave a fuck about foreign policy? Do you think they give a fuck about a plot of dirt next to some Jews? It's Muhammad's book that drives them. Fools will always look to blame their shortcomings on others; with September 11th, those guys thought they were dealing a blow to the Great Satan by starting a war. And now I hear a Muhammad stadium is being built around the corner from the first attacks.

MG: I believe so. But we have freedom of religion in this country and you can do what you want with private property.

SATAN: Really?

MG: Definitely.

SATAN: Can you build a strip club next to a daycare?

MG: (*seemingly perplexed*) Ugghh...

SATAN: Of course not. The community would object. The politicians would get involved and block it. However, the strip club would have to be allowed to open up somewhere appropriate and considerate to the communities wishes; perhaps just further north, outside the daycare district.

MG: But wouldn't a Muhammad stadium be a good thing for the area? Wouldn't it help foster understanding and a dialogue between faiths?

(*Satan spits out a bit of tobacco he finds on his lip*)

SATAN: Kid, if a bunch of assholes who didn't believe in Muhammad's religion, smashed a bunch of planes in and around, I dunno, Mecca... do you think that the people of the same faith as those assholes would have the balls to try and open up a stadium of their own... a couple of blocks away from crash sites... to foster understanding and a dialogue with the believers of Mohammad's religion?

(*Satan laughs*)

Kid, that stadium is political. How many imam's can raise the dough to needed to build in that location? None. It's a lot of dirty money invested to spread an ideology.

MG: But what if it's a community center?

SATAN: Calling a building with a Muhammad stadium on the first floor, or any floor for that matter, a community center is a laughable.

(*Satan produces an ashtray from his pocket and rolls the ember of his cigar in it*)

Kid, people over here are being shat on in the face... and they can't even smell it.

MG: But how can you say that. Surely there is some good will in the plans.

SATAN: You have to understand that we're dealing with a mindset that is outdated, expansionist and extremely patient. And time is on its side.

MG: Patient for what?

SATAN: For you to submit. In time, through conquest or destruction... you will submit. You just don't realize it.

MG: Why do so many people who believe in Muhammad's religion hate Jewish people?

SATAN: Money and Israel. Because Muhammad's believers are forbidden to charge interest, they have limited banking sectors, economic opportunities and poor savings among one another. If you can't make money off of lending money to a business or a person who needs money to fund something... where's the incentive to lend? How do you offset defaults? Are you meant to loan money in exchange for a return of a goat or a couple of chickens? Interest creates an incentive to facilitate and stimulate economic productivity. I'm surprised earning a profit of any kind isn't prohibited in Muhammad's book. Anyhow, many of Muhammad's believers see Jews as monopolizing banking and anything to do with interest.

MG: What about Israel? Surely there is no law against it.

SATAN: Oh there is. Infidels cannot rule any of Muhammad's land, it's written down somewhere I'm sure. And so they hate the place and the people residing within its borders. And they hate what they perceive to be an imposed suffering of Palestinians; but of course none of the actual believers in Mohammad's religion do anything to help them; if they did, which they could with all of the oil money, they'd have nothing to complain about.

(Satan extinguishes his cigar then pulls out a pack of cigarettes, removes one and lights it. He then burst out laughing)

Poor Jewish bastards. They should have set up shop somewhere safer and far, far away from all of the ignorance in that region. They have a right to be there and, I guess, they should be there. But if you buy a cottage... don't complain about the nature that surrounds you.

MG: You're losing me a bit.

SATAN: Sorry, the mushrooms are kicking in. Did I tell how much I couldn't really care about any of this?

MG: No.

(Satan slips awkwardly into the tub and grabs a bar of soap from a ledge)

SATAN: Well I don't. And fuck this bar of soap too! Fuck this bar of soap! That's hilarious! Normally the bar of soap is the reason you get fucked! Get it? Oh my god this is amazing! Wow! Did you just see that?

MG: Satan, you're high.

SATAN: And you're not. Stop being such a fucking buzz kill and eat these already.

(Satan hands Max Gold a bag of dried mushrooms)

MG: And then what?

SATAN: Write down everything that happened tonight.

MG: Really?

SATAN: Yeah really. Here, try this.

(Satan hands a computer over to Max Gold)

MG: I don't know about this?

SATAN: Sure you know! We've said it all already! Just write it all down.

MG: Ugh...

SATAN: And don't forget to throw in a cartoon of Muhammad.

MG: That's crazy. People will issue a fat twat on me just for dictating our conversion, let alone drawing a cartoon.

SATAN: Never give in to the irrationalism of a bully. These guys are bullies, kid. A person who tries to prevent an expression of will with the threat of violence… deserves no respect and can only be called on their bullshit if you engage them by exposing them for what they are.

MG: Yeah, but wouldn't a book on all of this be an intentional assault to the sensitivities of others?

SATAN: Then sensitive people shouldn't read it. It's a fucking book. People don't have to read it. You can't force people to think a certain way just like you can't force people to speak a certain way. People should question what they are spoon fed to believe from infancy. And it is right to comment and judge things, ideas, people and so on. Whether those comments and judgements have any merit… that is left for others to determine. It's what improves things.

MG: Ok. But I haven't drawn in years. It will look awful.

SATAN: Ah fuck it. Just give it your best. Listen, it's been grand, but I've gotta bounce. Good luck. Don't waste time praying if you have any questions. You have the answers already. Thanks for the bad wine and the shots.

MG: Don't mention it and, shit, thanks for the Caymus. It was great meeting you.

ACT V

~

(Satan exits through an extra dimensional portal of some kind. Max Gold eats the mushrooms, takes a long exhale and then reaches for the Apple)

~

MAX GOLD

ACT X

THE END

CPSIA information can be obtained
at www.ICGtesting.com
Printed in the USA
LVOW08*0345030817
543645LV00012B/240/P

9 780692 225004